The Amusing facts about chemical elements.

OGGN Publishing

Copyright © 2022 OGGN publishing.

All rights reserved.
No part of this book may be reproduced, or stored in a retrieval system, or transmitted in any form or by any means, electronic, mechanical, photocopying, recording, or otherwise, without express written permission of the publisher.

To A & K

&

all curious kids.

Contents

Title Page
Copyright
Dedication
Preface
Introductions — 1
Chapter 1 — 3
Chapter 2 — 21
Chapter 3 — 33
Chapter 4 — 51
Chapter 5 — 71
Glossary — 84

Preface

Dear reader,

This book will take you to the fascinating world of the elements. You will learn about and investigate the: most prevalent, noble gases, ancient metals, deadliest elements, and compounds.

A big part of this book is about **asking questions** and learning to **find answers** by observing things around you, and reading books and resources on the internet. So, what are you waiting for? The elements are waiting to be introduced.

Introductions
Who are the ELEMENTS?

The Elements are substances that are made up of only one type of atom. We will talk about atoms later in the book.

In the year 2021, there were 118 elements known to humans. On the next page, we have the 'family photograph' analogue of elements. Scientists call this visual as the *periodic table of elements*. Unlike a family photograph, though, the location of every element is fixed in the periodic table.

Periodic table of ELEMENTS

Chapter 1

The most prevalent

Quick question: what is something that is found in your **B**ody,

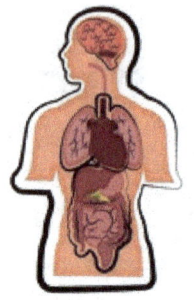

And in a **D**iamond ring?

Carbon!

Facts about Carbon

- ☐ Twenty percent of your body is made up of carbon!

- ☐ Both coal and diamond are forms of carbon.

- ☐ Carbon helps to make bubbles in soda and other *fizzy drinks*.

- ☐ Carbon is also found in stars and other planets. It is abundant in the universe!

- ☐ You might also find carbon in plastic, medicines, petroleum, baking soda, and even on the tip of your pencil!

Keep reading as **Seeker the slug** interviews **MR. CARBON**.

MR. CARBON

Seeker: Why are you everywhere, Mr. Carbon?

"Because I love to form bonds with other elements!' By the way, you can call me ^6C."

Seeker: What is the '6' beside the C, Mr. Carbon?

"The number 6 uniquely identifies me out of all other elements. The number 6 is my fingerprint but scientists call it atomic number. The atomic number determines an element's position in the periodic table. Go to the *periodic table* and find me on second row on the right-hand side."

Seeker: I have heard that you can be used to tell how old a fossil is. Is this true?

"Yup. I have another form called radioactive carbon. Radioactive carbon atoms decay at a fixed rate. In around 5,700 years only half of the original atoms of radioactive carbon will remain. In another 5,700 years only one fourth of the atoms of radioactive carbon remain and so on. The time taken for half of the atoms of a radioactive element to decay is called the half-life of the element. Using the half-life concept of the element, one can determine the age of the fossils."

Time to do some private investigations!

Find items at and around your house that have Carbon.

What is our SUN made up of?

Mostly Hydrogen!!

Interesting facts about Hydrogen

☐ The atomic number of Hydrogen is 1 and it appears first from the top-left in the *periodic table of elements*.

☐ Hydrogen is 14 times lighter than air. If you filled up a balloon with hydrogen and let it go, it will fly away.

☐ Water is actually made up of two atoms** of hydrogen and one atom of oxygen, H_2O.

(When you notice a word followed by ** in this book, please go to the **Glossary** section for an explanation.

☐ Hydrogen is highly **inflammable**, so we use another element HELIUM, a.k.a. ^{2}He in balloons, which is also lighter than air **but not inflammable.**

☐ Hydrogen exists as a gas at normal temperature.

☐ Liquid hydrogen was used by humans as a rocket fuel for space explorations, including the first human landing on the Moon!!!

You might have heard people talking about greenhouse gases. Methane is one of the greenhouse gases!

Methane is formed when one atom of carbon combines with one atom of hydrogen.

Compounds of carbon and hydrogen are called hydrocarbons. Examples of hydrocarbons include petroleum, and plastics, also called polyethylene.

(We will learn about COMPOUNDS in the last chapter)

Time to do some private investigations!

Josh heard his uncle saying to one of his friends that hydrogen will be used as fuel for aeroplanes by 2035. Is this true? Search on the internet and prepare a short report.

Why do apples turn brown when exposed to air?

because of exposure to
Oxygen!!!

Interesting facts about Oxygen

☐   Most of the oxygen in the earth's atmosphere came from plants via photo-synthesis.**

(** see the Glossary)

☐ We need oxygen for breathing.

☐ Even though we need oxygen to breathe, intake of too much oxygen can be deadly for us!

☐ At room temperature, oxygen occurs as a gas, O_2. The O_2 is the symbol of oxygen molecule**.

(** see the Glossary)

☐   In the Earth's upper atmosphere, we also find O_3 a.k.a. OZONE. An Ozone molecule has 3 atoms of oxygen chemically bonded together. Ozone protects us from the harmful radiation coming from the Sun.

☐   You need Oxygen for **fire** to happen, but Oxygen itself does not burn and is **not flammable**.

You might have heard about black holes. But, do you know what an OZONE HOLE is?

An Ozone hole is thinning of ozone in the upper atmosphere. This results in a rise in Ultraviolet (UV) radiation at ground level.

I bet that you know that too much exposure to UV radiation can cause skin cancer.

Time to do some private investigations!

Find out the main causes of ozone holes in our atmosphere and identify some ways that you can mitigate this process.

Cause	Remedies

Chapter 2

The Noble gases

Why are they called the noble gases?

The noble gases are so called because they hardly form chemical bonds with other elements.

They are so unlike Mr. Carbon who loves to form chemical bonds with other elements.

Which are the **noble** gases?

- → Helium (^{2}He)
- → Neon (^{10}Ne)
- → Argon (^{18}Ar)
- → Krypton (^{36}Kr)
- → Xenon (^{54}Xe)

→ Radon (^{86}Rn)
→ Oganesson (^{118}Og)

Interesting facts about Noble Gases

☐ **Helium** got its name from *Helios*. In Greek mythology, Helios is the sun god. Scientists had first detected Helium in the sun, and so it was names *Helium*.

☐ Did you know that Neon is responsible for the bright street lights and advertising panels?

☐ We love our double-glazed windows, don't we? They keep us warm in winters. But, did you know that there might be **Argon** in between the glasses?

☐ Argon is a bad **conductor**** of heat and so putting argon between the glasses prevents heat from escaping through our windows.

What is the meaning of the term 'cryptic'?

(If you do not know the answer, feel free to consult a lexicon also known as a dictionary)

☐ **Krypton** (^{36}Kr) got its name because it is very rare in nature. In Greek 'kryptos' means "hidden".

☐ Krypton is used for making **fluorescent** lights.

☐ Krypton is also used as a flash lamp for high-speed photography.

Xenon (^{54}Xe) is another interesting member of the noble gases. Unlike other noble gases it **CAN** form compounds** with some elements.

- ☐ Xenon is used in the car headlights. If you see car headlights that emit a soft blue glow, they are most likely using xenon.

- ☐ In the olden times, if you needed a surgery to save your life, you would have had to tolerate the pain as the surgeon cut you open and then stitched you back. That really must have hurt!!!!

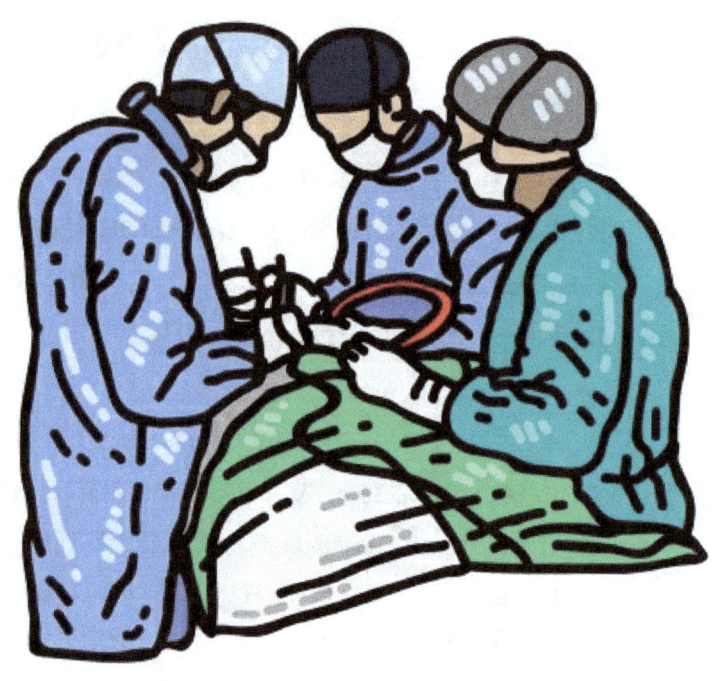

☐   Fortunately, we have good anaesthetics now and you won't feel any pain. Phew!

☐   Xe is also used as an anaesthetic**. But, because it is a rare element, Xe anaesthesia is expensive.

Oganesson is a SYNTHETIC element.

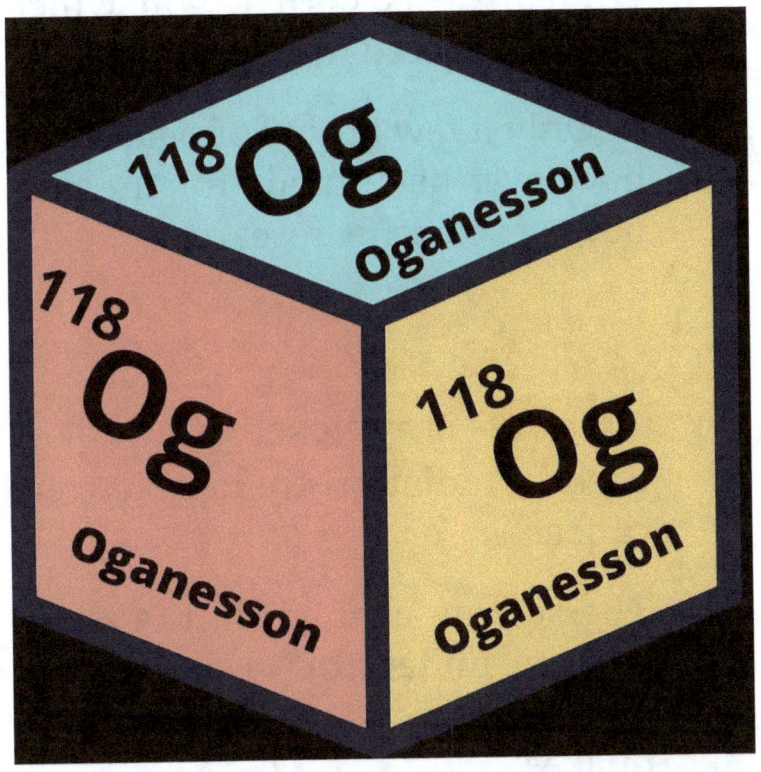

☐   Synthetic elements were not discovered occurring in nature but have been artificially produced by humans in labs. Yay!

- ☐ Oganesson is also radioactive!
- ☐ It was discovered only in 2002. Compare this with Argon which was discovered in the 1890s.
- ☐ Only a few atoms of oganesson have ever been made for the purpose of scientific research.

Radon is radioactive and it is used in radiation therapy to treat cancers.

The sudden changes in the quantity of underground radon can also be used to predict earthquakes.

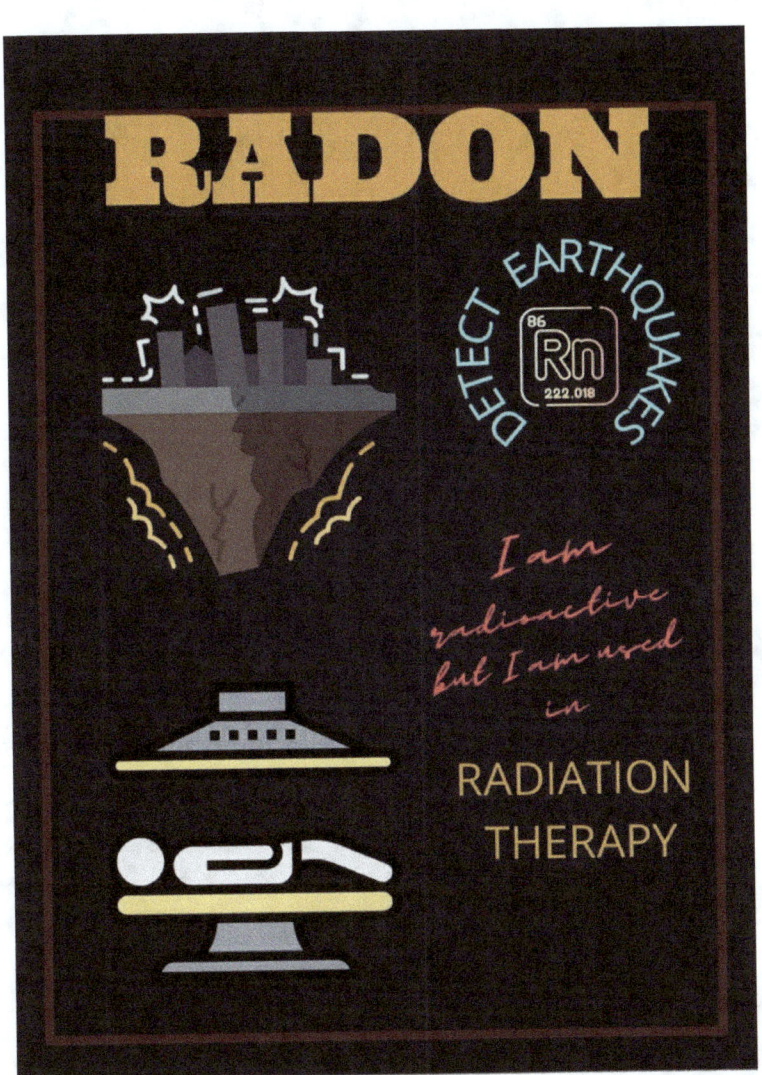

Time to do some private investigations kids!

Grab a A4-sized paper. Create a poster on any of the NOBLE gases discussed in the chapter. You can refer to the poster on radon for inspiration.

OR

Ask a grown-up in your household whether they have ever used *anesthesia* in their life. Did they know what type of anesthesia it was?

Chapter 3

The ancient elements

The discovery of metals was so significant in human history that an entire era is named as the Bronze age and Iron age.

Unlike stones, metals could be melted and moulded into any shape.

What were some of the metals used by cavemen? What did they use metals for?

- ☐ Some of the metals used from the ancient times include gold, silver, copper, tin, lead, iron and mercury.

- ☐ Cavemen used metals to make weapons to defend themselves and for hunting.

- ☐ They might also have used metals for making jewellery.

Let us learn about some of these metals. The study of metals is called **METALLURGY.**

Gold

Gold can occur naturally in nature. There are many rivers, and lakes around the world where you can find gold particles glittering with the sand particles. It is quite possible that gold was known to humans from prehistorical times. Pre-history is that period of human history where there are no written records.

Egyptians were perhaps the first civilisation to use gold extensively.

You might have heard about Pharaoh (King) Tutankhamun. Tutankhamun was only 19 when he died. Historians have found over 100 kilograms/200 pounds of gold in his tomb!

Gold was used for making jewellery. Gold was also used for making diadems, amulets, ornamental weapons, and vessels etc.

Gold was also used in the ancient times to make coins (currency). King of Lydia, Croesus, introduced gold coin as a currency/money.

A GUESSING GAME

Today the three of the biggest producers of gold are:

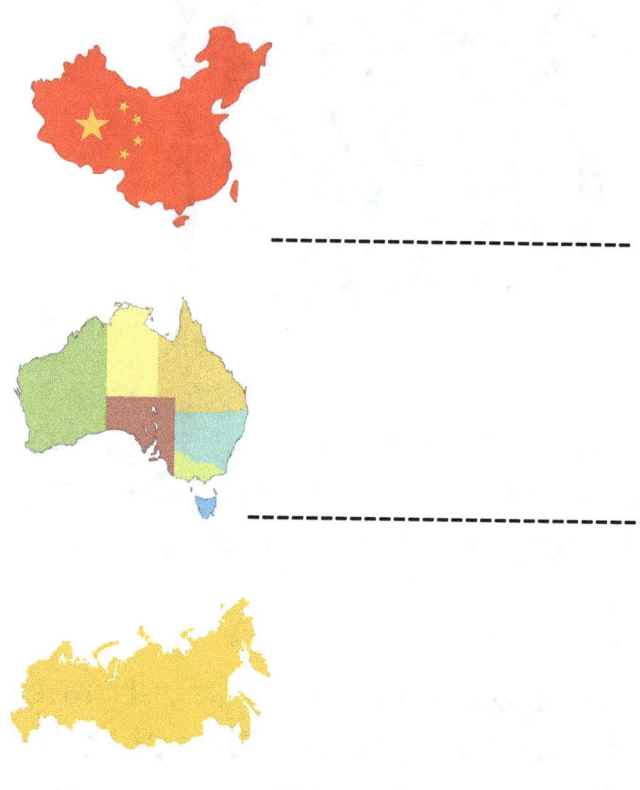

_____

_____

_____

ANAGRAMS

Some of the biggest producers are also the biggest consumers of gold.

Can you guess countries that are among the biggest consumers of gold? Hint: these are not the countries listed above as the biggest producers of gold.

Today, some of the three biggest CONSUMERS of gold are:

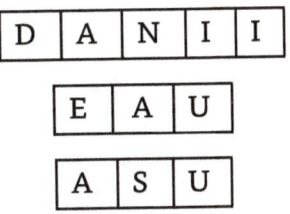

Solutions to guessing game
CHINA, AUSTRALIA, RUSSIA
INDIA, UAE, USA

What makes gold so special?

☐ Gold does not lose its glitter for hundreds of years.

☐ It does not react with other elements easily.

☐ Gold is one of the most ductile** and malleable** elements.

☐ Gold is a good conductor of heat and electricity.

☐ It is rare in nature.

☐ Today, aside from making jewellery, and bullions, gold is also used for making medals, in electronics, in space vehicles, and in tooth fillings.

☐ The scientific name of gold is Aurum and its atomic number is 79.

Bronze

Bronze is actually an alloy of copper and tin. It is yellowish-brown in colour.

Bronze is used for making medals, coins, decorative items, automobile parts, and utensils.

Copper/Cuprum in Latin (^{29}Cu)

Copper is widely used to make wires for electrical transmission. It is a good conductor of heat and electricity. Copper is a good conductor of heat and electricity.

Copper is **reddish brown** in colour but

when it comes in contact with CO_2 it reacts and turns green.

You might also find copper utensils at home or in the door knobs.

Time to do some private investigations kids!

Can you locate any other place/items in your home that has copper?

Red blood and blue blood

☐ Most animals including humans have red blood. The red colour comes from haemoglobin, a protein that has **iron** within it.

☐ But, did you know that Crustaceans, spiders, squid, octopuses have blue blood because their blood has haemocyanin?

☐ Haemocyanin has copper instead of iron.

☐ Scientists have found that some creates also have green and violet blood! **Why don't you find out what makes their blood green and violet?**

- ☐ Iron (^{26}Fe) is also called Ferrum.
- ☐ Rust is formed when iron reacts with oxygen and water.

☐ Iron is one of the most abundant element on Earth as well as in many meteorites.

Fun time
Time to do private investigations!

☞Create a poster on silver.

☞Do some research on tin over the internet or in your local library and identify at least 1 item made of tin in your house.

☞Find out which creatures have green and violet blood and why.

☞Find 5 words made up of 3 or more letters with using the letters of the word **METALLURGY.**

Solutions to the word finder.

TALL
MALL
LEG
RUG
YAM
RAGE
GALE
GALL
MALE
LAME
MEATY
TALLER
LATELY

(There are other solutions as well!)

Chapter 4

The deadliest elements

Game time!
I SPY ELEMENTS

H								
Li	Be							
Na	Mg							
K	Ca	Sc	Ti	V	Cr	Mn	Fe	Co
Rb	Sr	Y	Zr	Nb	Mo	Tc	Ru	Rh
Cs	Ba		Hf	Ta	W	Re	Os	Ir
Fr	Ra		Rf	Db	Sg	Bh	Hs	Mt

	La	Ce	Pr	Nd	Pm	Sm	Eu
	Ac	Th	Pa	U	Np	Pu	Am

PERIODIC TABLE OF

I SPY ELEMENTS

								He
			B	C	N	O	F	Ne
			Al	Si	P	S	Cl	Ar
Ni	Cu	Zn	Ga	Ge	As	Se	Br	Kr
Pd	Ag	Cd	In	Sn	Sb	Te	I	Xe
Pt	Au	Hg	Ti	Pb	Bi	Po	At	Rn
Ds	Rg	Cn	Nh	Fl	Mc	Lv	Ts	Og

Gd	Tb	Dy	Ho	Er	Tm	Yb	Lu
Cm	Bk	Cf	Es	Fm	Md	No	Lr

ELEMENTS

Locate the following elements in the periodic table:

^{33}Ar *a.k.a.* **Arsenic**
^{82}Pb *a.k.a.* **Plumbum**
^{80}Hg *a.k.a.* Hydrargyrum, commonly known **Mercury**
^{84}Po *a.k.a.* **Polonium**
^{87}Fr *a.k.a.* **Francium**

Even though human body is made up of elements, some elements can be **deadly** for us! Don't believe me? Then, read on.

The Arsenic Story

Question: what can Nitai do to get better?

***Time to tie the loose ends
and find the culprits***

Arsenic occurs naturally in the groundwater in a number of countries. When people drink this water, they can get arsenic poisoning.

According to scientists, around 500 million people world-wide are exposed to water contaminated with arsenic.

Arsenic is tasteless and odourless. You can be exposed to it without your knowledge.

Symptoms of arsenic poisoning

☐   Short-term exposure might result in abdominal pain, nausea, and abnormal heart rhythm

☐   Long-term exposure can result in skin lesions and cancer.

Prevention: Drink and cook in clean, filtered water. If diagnosed with arsenic poisoning seek medical attention.

Other facts about Arsenic

Arsenic is a metalloid. Refer to the **glossary** for a description of metals and non-metals.

Arsenic was discovered in 1250 by Albertus Magnus.

Arsenic is used in the electronics industry and in the production of wood preservatives.

Plumbum, ^{82}Pb

In Latin, plumbum means 'waterworks'. In ancient Rome, lead was used to make water pipes. Hence, lead is called plumbum.

Where might have the word plumber originated?

Lead is present everywhere. It can be the pipes in our homes or in the paints on our walls, ceramics (pottery) and toys. It might be present in gasoline, batteries, ammunition and even in cosmetics.

Lead can get into your body when you consume contaminated water or food, or when you breathe fumes or dust that contain lead.

There is **no** level of lead that is safe for human body. In children, even slightly elevated levels of lead in the blood can cause learning disabilities, seizures, and even death.

Exercise: Look at the ingredients section of the make-up items used in your household to find out whether they have lead.

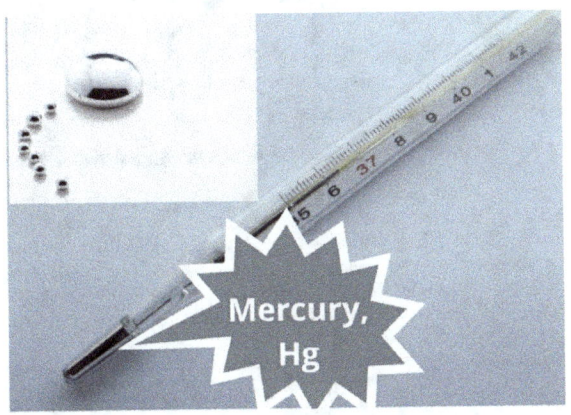

Mercury, ^{80}Hg (hydrargyrum) got its name from the Roman god Mercury, who was known for his quickness. Mercury is also known as quick-silver. Can you guess why? Hydrargyrum means liquid-silver in Greek.

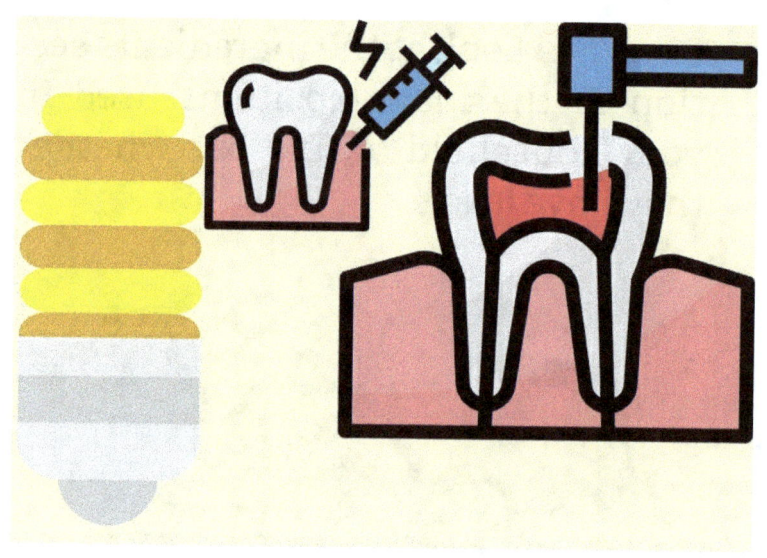

→ Mercury was known to mankind in ancient India and China.

→ Mercury and its compounds are highly toxic to humans.

→ You might have seen mercury in a thermometer. It is also found in fluorescent lamps, and dental amalgams. Dental amalgam is a dental filling material used to fill cavities caused by tooth decay.

→ Some mercury compounds might be found in contact lens solutions and cosmetics.

Polonium

→ ^{84}Po, Polonium is a highly radioactive element.

→ It was used as a trigger in the atomic bombs, including the ones dropped on Hiroshima & Nagasaki. Polonium was discovered by a scientist named Marie Curie in 1940. It was named after Curie's home country, Poland.

Enjoy this short comic based off of an actual event in Irene Curie's lab. Irene was Marie Curie's daughter.

→ Polonium has also been used to poison people. It is so lethal that 1 gm of Polonium can kill millions of people.

→ In order to murder someone by polonium poisoning, the victim would have either inhale or ingest the element.

→ Some of the symptoms include: nausea and vomiting, hair-loss, lowered white blood cell count, and damage to bone marrow.

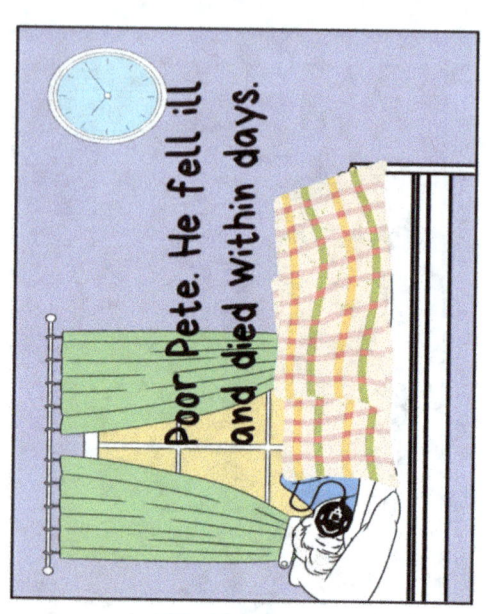

FRANCIUM

You have guessed it right! **Francium,** ^{87}Fr was named after France.

Marguerite Perey discovered it in 1939 while trying to create the purest form of another element, Actinium.

FRANCIUM is the heaviest of **alkali** metals. It has biggest atom of any element in the periodic table.

Alkalies react with acids** to produce neutral salts. Alkalis are corrosive to organic tissues.

Francium is highly reactive with water. It can react with even the moisture in your hands. It is also highly radioactive.

Francium is the second rarest element in the Earth's crust. It has a half-life of only 22 minutes.

Private investigation!

What is the **half-life** of ^{238}**U?**

Chapter 5

Compounds

Compounds are molecules made up of two or more atoms of different elements.

A molecule is made up of one or more atoms.

Examples:

O_2 = Oxygen (a molecule)

O_3 = Ozone (a molecule)

CO_2 = Carbon dioxide (a compound)

H_2O = Water (a compound)

NaCl = Table salt (a compound)

H_2S = Hydrogen sulphide-a compound (sewer gas, smells like rotten eggs)

H_2O_2 = Hydrogen peroxide-a compound (used for bleaching among other uses)

$NaHCO_3$ = Baking soda-a compound

Molecules and compounds are formed when two or more atoms are chemically bonded together.

What do you see?

The picture actually shows two most common compounds on our planet, water (H_2O) and sand (SiO_2).

Organic compounds always have CARBON. Usually, organic compounds also have hydrogen and oxygen chemically bonded with carbon.

Most inorganic compounds don't have carbon.

Our body and organs are made up of organic compounds.

Our body needs water to function properly, which is an inorganic compound.

Does your toothpaste have fluoride?

Take a look at the cover of your toothpaste to find out whether it has fluoride?

Compounds having Fluoride are great for preventing tooth-decay in children.

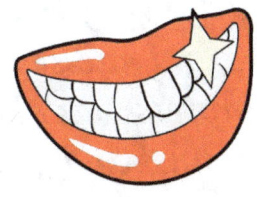

If you use a natural toothpaste, your toothpaste might have calcium carbonate and sodium bicarbonate. Sodium bicarbonate is also called baking soda. Very useful compounds huh!

Did you know that calcium carbonate is also called white marble? Yeah! the beautiful white stone that you might find in your floor or kitchen/bathroom tiles and famous monuments.

STORY TIME !

Once upon a time there lived an emperor in India who loved his wife, the empress, dearly. They lived happily for several years and had several children. When the empress was about to die, she made her husband promise that he would build a tomb that was unlike any other in the world.

After the death of his beloved wife, the emperor built the Taj Mahal in her memory. It is listed as one of the seven wonders of the world. The Taj Mahal is made up of white marble!

Your favourite chocolate and ice-cream are likely to contain sugar which is also a compound.

Proteins such as eggs and meat are complex organic compounds.

Let us go compound hunting around the house.

> Soap

> Mouthwash

> Paint remover (acetone)

- Vinegar (an acidic compound)
- Car battery acid
- Petrol/gasoline
- Plastics

➤ Dress materials such as cotton and wool.

Time to do more private investigations!

Find some more compounds to finish the list on the previous page.

What are your **top 3** favorite compounds?

This Brings Us To The End Of The Short Intro To The Elements Family...

Did you know that humanity has not discovered all the elements that exist in this universe? Perhaps, you will discover a new member of the **ELEMENTS** family.

Glossary

Acid
Acids are corrosive compounds. Examples include citric acid found in lime juice and vinegar.

Alkali
Alkalis are chemical opposite of acids. They dissolve in water and neutralise acids. Examples include baking powder, and indigestion tablets.

Anaesthetic
A substance that induces insensitivity to pain.

Alloy
A mixture of two or more metals such as steel and brass. Steel is an alloy made up of iron, carbon and other elements. Brass is an alloy made up of copper and zinc.

Atom
An atom is the smallest building block that has the features of an element. Atoms are made up protons, electrons, and neutrons.

Bullion
A rectangular slab of pure gold/silver.

Base
A base is a substance that reacts with acids to form salts. Example, baking soda.

Compound
Compounds are molecules made up of two or more atoms of different elements.

Conductors
The property of a substance to allow heat and electricity to pass through it is called conductivity. The substances having this property are called conduct-

ors.

Ductile
The property of an element to be able to be drawn into thin wires.

Element
A substance made up of only one type of atom.

Malleable
To be able to hammer something into the desired shape without cracking it.

Metals
Metals are substances that reflect light easily (shiny), can be shaped easily by hammering or forging, and through which electricity and heat can pass easily.

Metalloids
Elements whose properties lie in between those of metals and non-metals.

Molecule

A molecule is made up of one or more atoms such that they act as one particle. For example, oxygen (O_2) molecule and ozone (O_3) molecule are both made up of element oxygen. Their properties are however, very different.

Non-metals
Non-metals are opposite of metals. They cannot be shaped easily, are not shiny, and are poor conductors of heat and electricity.

Photosynthesis
Photosynthesis is the process by which plants use sunlight, carbon dioxide-CO_2, and water-H_2O to produce nutrients for them. Oxygen, O_2, is released in this process.

Radioactivity
The emission of nuclear radiation.

THANK YOU VERY MUCH FOR READING!!

We would like to acknowledge Britannica.com, Royal Society of Chemistry, and numerous other resources on chemistry that have made the compilation of this book possible.